Rethinking Math Learning

Rethinking
Math
Learning

Teach Your Kids 1 Year of Mathematics in 3 Months

Dr. Aditya Nagrath, PhD

HOUNDSTOOTH
PRESS

RETHINKING MATH LEARNING

Teach Your Kids 1 Year of Mathematics in 3 Months

ISBN   978-1-5445-1520-5 *Paperback*

978-1-5445-1519-9 *Ebook*

For my son, Elliott—I love you

Contents

Introduction

___

Whether you have a doctorate in mathematics or only took the bare minimum math prerequisites to get your diploma, math is something we all experience on a regular basis. For many people, math provokes anxiety. That anxiety may have affected your career choice, altered the course of your life, or caused you to simply believe that you're "not a numbers person."

Now as a parent, you find yourself back in math class as you help your child with their homework, only to find that the curriculum has changed from when you were a kid and you need to learn the subject anew. You may even find yourself confronting your own math anxiety all over again.

The good news is, being involved with your child's

math education is the first step toward their success. Every teacher and research study will point out that academic outcomes are significantly better when a child has a parent who is involved in their education. For me, that was my mother.

Every summer, my mother would gather our math books for the upcoming school year. She would sit down with me and my sister and ensure that we understood the concepts. This was not always an easy road, and tears at math time are something I deeply recognize and understand. Her efforts played no small part in shaping the course of my life. I graduated with a doctorate in mathematics and computer science just as the 2008 financial crisis altered life in America, an event partially caused by erroneous mathematics.

However, the science of education has progressed so much in the last forty years that scientists now know exactly how children learn and understand this subject. What comes as a surprise to many people? Memorization is no longer the key to being able to succeed in math.

It's common practice to help children learn math

by memorizing multiplication tables and formulas. This helps children get good grades on their tests and pass their math classes. Memorization only gets a child so far, though. It's crucial to understand early on that once you get to algebra and beyond, math becomes a language; everything that came before was basic vocabulary to prepare for more abstract conversations that occur in that language. It is an endless jargon that builds upon itself—not unlike the jargon that goes along with any profession.

So why are we still teaching children to approach math with memorization?

What I have learned as both a mathematician and as a father is that your child's success in math is partly about retooling the way we *teach* mathematics and partly about reframing the way we *think* and *talk about* mathematics.

To help your child succeed in math on a long-term basis, you must change the way you—as a parent, grandparent, guardian, or teacher—view math's role in your life and your child's life.

For many decades, mathematics mattered most to

scientists, engineers, technologists, and doctors. But with technology's mind-boggling growth, mathematics is no longer relegated to science-specific careers and industries. While STEM-based careers still produce the best-paying, fastest-growing jobs, the reality is, whatever field your child ends up going into, they will need math in some way. Mathematics appears as a major function of seemingly unrelated careers, from marketing and graphic design to skilled trades such as plumbing and electrical work. Rarely do the professionals know they are exercising mathematics as they exhibit the concepts at play in their daily lives.

Even if your child becomes one of the few people who never needs to understand the numbers on the job, math gives children confidence and skills to perform well in other areas of life. In fact, one study showed that preschool math scores predicted overall academic performance in fifth grade! That means children did better in all subjects, not just math, if they established a solid math foundation in preschool. Confidence in mathematics helps develop problem-solving skills and intuition around numbers and logic. That's because math works the brain like a muscle, developing logic, reasoning, and problem-solving skills, as well as overall mental acuity. Research shows that children who do more mathematics are better readers, writers, and problem solvers.

This doesn't even touch the extended benefits of a math-confident society. People in STEM careers are the change makers. These are the people who will solve global issues that affect generations to come—climate change, hunger, economic inequity, and so on. Imagine what we could accomplish if the next generation of children grew up with math confidence. It would produce both the scientists who could solve global issues *as well as* businesspeople and politicians who would understand the language and enact change.

Once you see the importance of math and how it affects your child and society as a whole, you can approach math homework differently. Instead of being something to suffer through, math becomes a valuable tool that can put your child ahead of the pack in whatever career they choose.

As you realize this, you'll also realize that not all math curricula are equal. That's where Elephant

Learning differs and why we teach math the way we do. We didn't simply develop a fun computer game for kids and add math problems to the mix. Instead, we *gamified a proven math curriculum*. We made solving math problems part of a fun puzzle instead of a dreaded task.

The result is the most effective tool for teaching mathematics ever created. We have found that on average, children in our system learn one and a half years of mathematics over the course of ten weeks when they use our system an average of thirty minutes per week.

We didn't stop there. We continuously strive to improve our system. We analyze data on common pitfalls and rework our program behind the scenes to ensure children are always getting the best in-app experience. At the same time, we empower parents through reporting and understanding. If we can show you, the parent, how we intend to accomplish our goal of teaching your child a certain math concept, then you're empowered to help accomplish that goal.

I firmly believe there is no such thing as being

"not a numbers person." I believe any person can accomplish anything with awareness and adaptation. The belief that we cannot achieve stems from small gaps in understanding. Fortunately, we can close those gaps.

Math is a powerful tool that propels your child to succeed in school and throughout their entire life. You may find as your child overcomes their anxiety toward mathematics, they are able to overcome any other obstacle that stands in the way of their wants and dreams.

—DR. ADITYA NAGRATH

About This Book

—

Maybe you want to help your child with their math homework or accelerate their progress. Maybe you're a homeschooling family looking to expand your math curriculum. Perhaps you're part of a school or other learning institution wanting a customizable, scalable mathematics enrichment program. Whatever the goal, this book and the Elephant Learning app can help you achieve it.

This book will break down math concepts the Elephant Learning way. We'll begin by showing you the increasingly important role of math in everyday life. Then we'll dive into the ages and stages of math learning, beginning with the toddler years all the way up to algebra and beyond. You'll walk away equipped with actionable methods of incorporating math into your child's daily life in a way that is not only natural but also produces results.

Chapter One

The Real Reason Math Curricula Are Failing Your Child

———

Does your child enjoy math class? When you see them doing their math homework, does it feel like they don't really get the concepts? Do they appear to blindly apply strategies they've been taught in class to solve their homework problems?

In the classroom, many children are unable to develop a solid math foundation due to the typical way math is taught. The good news is, you can remedy this issue at home by simply looking at math instruction through a new lens so that your child goes into the classroom prepared to take on mathematical challenges.

INSTRUCTION VERSUS EXPERIENCE

Teachers are accustomed to teaching mathematics through instruction. It's not that this strategy is incorrect; it's simply the most practical strategy to employ when standing in front of a classroom full of students.

When a student doesn't get a math concept, a teacher may then instruct the students on "how" to solve a problem with a step-by-step procedure to memorize and use.

The issue with this is, strategies are better discovered than memorized. If your child simply memorizes a strategy, can you be sure they truly understand the concept and language, even if they can get the right answer?

Think of it this way: You can't really instruct a child on what the color red is. You can show a child red objects and you can label them as red, but you can't necessarily tell them what red is. Even if you read the definition of "red" in the dictionary, your child still won't understand what red is without seeing and experiencing the color for themselves.

In the same way, how do you describe addition to a

child? "5 + 4 means 'Give me five objects. Give me four more objects. Now how many do I have?'" If a student has not had the experience of this simple activity, the only thing that can be done besides going back to ensure they understand the fundamentals is to memorize the answers. After all, there is a test coming up!

Imagine walking into a third-year lecture in organic biochemistry (or if you are a biochemist, a third-year lecture in graduate mathematics). The lecture is full of jargon. One university student I know described it this way: "It sounds like they're speaking English, but I have no idea what they're saying!"

This is what three out of four elementary students experience in math class. Children are being tested on materials they don't understand. Memorization is the only strategy that appears to work!

Eventually, this process will fail. If they did not understand the prior math concepts, memorization no longer works as a strategy for passing homework and tests when children get into more advanced mathematics curriculum such as algebra. Algebra is the first topic that students encounter where we start having conversations in mathematics, using the elementary concepts built up to that point.

For many parents, they never understand that this—the mere memorization of procedures to solve problems without any understanding to back it up—is what's happening with their child. They have no idea what their child is going through in the classroom. As parents, if we ask the student, "What is 5 + 4?" and they tell us "9," that is typically good enough. For most teachers, that is also good enough. It sounds correct, and time is often too limited to check further.

Children learn mathematics through problem solving, logic, and reasoning. Just like with colors, the best way to help children understand math is by giving them mathematical experiences of each idea and then placing the language around it. This must be done at the student's level of understanding to be effective.

By learning mathematics through experience, your child discovers strategies and how to use them to solve real-life problems, rather than just memorizing tables. This is how they build intuition and problem-solving skills.

SETTING YOUR CHILD UP FOR SUCCESS

We can't blame this issue entirely on the school system and teachers. Research shows if children come into kindergarten understanding mathematical concepts, then the US school system produces great students.

That's where working with your child at home gives them a huge advantage. In nearly every study on education, outcomes are vastly improved when parents are involved in the learning process.

Being able to effectively teach my child mathematics at home is the reason I created the Elephant Learning platform. Not only does it simultaneously teach and evaluate, but it also provides valuable feedback to parents on how to make further progress outside of the app with fun activities, such as board games. What we hope to do in this book is to teach you our approach so that you as a parent can replicate our methods at home.

Helping your child understand math concepts at home is not about instruction or showing them how to solve problems. For example, the Elephant Learning app does not "instruct." We define and we give students mathematical experiences that help them comprehend math concepts. It is important to distinguish between showing a student how to get a solution and ensuring that they understand the question and the ideas. If a student memorizes how to get a solution, when testing them later it will be challenging to determine whether they understand the topic or are simply regurgitating the solution or procedure you showed them.

The principle goes back to the concept of teaching "red." It's giving the child the experiences of red versus giving them a definition of red that helps them truly understand what the color is and how to recognize it.

When parents use Elephant Learning as directed, we receive testimonials raving about how their children have become more confident. They do better on tests and actually enjoy math class, because they finally understand the teacher in the classroom.

Without this kind of support, children with math anxiety unfortunately become adults with math anxiety. Since I started Elephant Learning, so many parents have approached me with stories of how they wanted to be a physicist or engineer but did not have the desire to engage in mathematics any longer. It is heartbreaking to hear about people giving up on their dreams due to a fear of mathematics. By the end of this book, I hope everyone will see that mathematics is not something to be anxious about; rather, it is a set of powerful tools that, when understood, allow you to solve real-world problems.

Chapter Two

Math Anxiety

When I was young, every summer my mother would sit me and my sister down to learn the math concepts for the upcoming school year. Math time at our house wasn't always a calm time. There were definitely some tears, and I'm sure it felt like pulling teeth for my mother.

Beyond our not really wanting to do math during summer vacation, there was also a fear of getting these new math concepts and problems wrong—in other words, math anxiety.

The next thing I knew, I was in fourth grade, and I didn't get into the advanced math class by one point. Still, I was fighting to get in. My math anxiety drove that competitiveness; I couldn't, for whatever reason, *not* get into the class. I pushed to get ahead of my peers, learning the advanced math concepts. I also learned *how to learn* from this experience, a valuable life skill I would employ throughout my life and career to "get to the next level."

I ended up with a doctorate degree in mathematics and computer science. Until I understood what math anxiety was, I didn't realize that it was the anxiety driving me.

From there, I could see how it affected me and how it affects others every day. Over the last four years, standing on the shoulders of the best researchers in the science of mathematical education, I've examined the self-perpetuating cycle of not understanding mathematics leading to anxiety, and anxiety leading to not understanding mathematics. We designed Elephant Learning to help students overcome this problem, and by the end of this book, you will understand the techniques we use so that you may help your student overcome math anxiety or never develop it in the first place.

HOW CAN I TELL IF MY CHILD HAS MATH ANXIETY?

The first step in addressing math anxiety is to recognize what it looks like.

When you and your child sit down to do their math homework at night, do they experience any of the following?

- Tantrums and tears
- Frustration
- Fear and dread
- Low self-esteem or low confidence
- Anxiety

The first step in addressing math anxiety is to recognize what it looks like.

Are they emotional about answering math questions incorrectly? What are their emotions around poor math grades (if they have them)?

How do they deal with a word problem? If you ask them a word problem and they don't understand the question or are unable to figure out which tool (addition, subtraction, multiplication, etc.) to use in order to solve it, then this *may* be a good indicator of some math anxiety.

Word problems are ways we test students' understanding of the ideas. For example, in Elephant Learning, to test multiplication, we set up a problem where there are seven objects displayed per row in a grid, and there are seven rows displayed. The student has a guide at the fifth row, so they only need to be able to eyeball that there are two more rows than five. We show them this problem for four seconds and then ask how many objects were displayed. To answer this question, the student needs to know their multiplication tables to multiply 7×7—and they also need to know that using multiplication is the way to solve the problem.

If a student does not understand the ideas, imagine how math class looks to them. They do not understand the teacher; they are memorizing what feels like arbitrary information through multiplication tables and methods of adding and subtracting multiple-digit numbers. They do not understand why they are doing it. In fact, the most common question math teachers receive from students is, "When am I ever going to use this?" The answer to that question, sadly, is the following: if you do not understand the idea behind the tool, you will likely never use the tool.

This situation—combined with the stress of grades, exams, quizzes, and parents wagering the student's future on academic performance—is a lot, especially for young students who may not have learned appropriate coping skills. The result is fear and self-defeating beliefs, followed by outbursts, low confidence, frustration, inability to hear (they are listening, but they are not hearing you), or other symptoms of math anxiety. They all fall on the spectrum of flight-or-fight response.

Try this exercise: Grab a handful of the LEGOs that have four dots on top. Ask your child to give you five of those blocks and tell you how many dots there are in all. If your child counts to get the answer, don't worry. Let them come up with the answer.

Ask them next, "What is 4×5?" This may help them connect the concept of multiplication to the memorized times tables. However, if you try an exercise like this and find your student counting, it typically indicates they have a gap in understanding the concepts and would be prone to developing math anxiety.

HOW CAN I HELP MY CHILD COMBAT MATH ANXIETY?

Help your child combat math anxiety by filling in the gaps in understanding, starting at their level. As they see success, confidence builds. That is why, within the Elephant Learning platform, the Elephant Age is front and center for the student. As they begin experiencing success, it is always measured and displayed to them, priming them to want more.

As a strategy, start giving your child math experiences by incorporating concepts into real-world situations. People, including children, get dopamine when they solve problems, more so when those are real-world problems or help them gain insight into the world around them. It is important to do this at your student's level of understanding. So, for example, if your student is learning to count, then have them count stairs, cars, chairs, and so on. Or ask them to put out three forks. Now, when math time comes, if it is also at their level, they view it as a way to solve problems they are already familiar with and see every day around them.

This is exactly what Elephant Learning does. The entire system was built to remove math anxiety and facilitate the learning of mathematics. The app's games are educational on their own, but if your child gets stuck, the app provides you with a series of questions to ask that will help your student overcome their hurdles.

Once you've filled the gaps in understanding, the narrative of "I'm not good at math" will no longer reflect reality, and you may find increased confidence. It is also important to understand the story and help children be aware these beliefs do not reflect reality. It is a good opportunity to quote Henry Ford: "Whether you think you can or can't, you are right." It is important for students to understand they can accomplish anything they want, especially when it comes to mathematics. It is just a matter of learning the jargon and practice.

HOW DO I PREVENT MY CHILD FROM HAVING MATH ANXIETY IN THE FUTURE?

There are a few things you can do to keep math anxiety at bay in the future:

1. **Keep math fun.** Make it a game and always be playful around math.
2. **Teach at their level.** When you're talking over your child's head, they can start to feel anxious again.
3. **Be mature about your own possible math anxiety.** When you start to get burned out, it's okay to step back and take a break.

The Elephant Learning app does this, too. It explains how it's teaching the subject and why, then breaks the topic down further into milestones. Parents can find activities to do with their child outside of the system that teach the same concepts so the child receives more exposure to a concept to learn it faster.

FINAL FACTS ABOUT MATH ANXIETY

Make no mistake, math anxiety can and does affect the course of your life. The wife of a friend said to me, "I wanted to get a degree in physics, but it was all differential equations, so I became an English major." When she was a child, she wanted to be a physicist, but she gave that dream up because of math anxiety.

Creating beliefs such as "I am not a numbers person" becomes a self-fulfilling prophecy. Whether it's you or your child, those who are not confident with mathematics are typically individuals who have math anxiety.

Regardless of how much math anxiety exists in your household, remember: there is a solution, and that solution can play a large role in your child's future success.

Chapter Three

Why Children Are behind in Mathematics

———

The majority of children are behind in mathematics. There is no one to blame here, especially not the parents. As we have already learned, students who have adopted the memorization strategy may seem indistinguishable from those who understand mathematics, even to trained teachers.

MATH DEFICIENCIES START IN KINDERGARTEN

Research shows that four out of five students enter kindergarten unprepared for the curriculum. Kindergarten mathematics curriculum requires a student to understand how to count to ten upon entering. For the school, counting to ten means that the child can pick ten items out of a pile and give them to you when you ask for them—no more, no less. This is challenging for young students because they must hold the number in their head to remember to stop on that number.

Prior to starting Elephant Learning, I would have thought counting to ten meant my child could verbally recite the numbers one through ten. I would have asked my child to do that and sent them to kindergarten thinking they were prepared. This is where most parents are being let down and how the cycle of math anxiety begins.

If a child enters kindergarten knowing how to count to ten by the school's definition, they tend to do fairly well within our education system. Unfortunately, only the top 20 percent of income earners are preparing their children for kindergarten in this way, mainly because they can afford to send their child to preschool, where children learn how to count to ten by this definition. The other 80 percent of our children enter into a system that passes them along to the next grade level regardless of their level of understanding.

PRESCHOOL MATHEMATICS CAN PREDICT THE REST OF YOUR CHILD'S LIFE

This early math readiness is significant because the research also shows that at the preschool level, children who do more math are better readers, writers, and problem solvers. They have better grammar and better reading comprehension.

For example, one study shows that preschool math scores are a better predictor of third-grade reading scores than preschool reading scores, meaning the more math children do, the better they are at reading down the road.

Another study shows that preschool math scores predict fifth-grade *overall* scores, not just fifth-grade math scores.

Analyzing this last study, I believe there are two reasons why this could be true, and I am sure some combination of the two factors are at play for any particular student.

First, math tends to be like mental gymnastics: it exercises your mind. Children who are doing more math are practicing mental skills more often. They are exercising problem solving, working memory, and more. Just like walking and chewing gum at the same time requires more practice, counting and remembering what you are counting to is a more challenging skill for young students.

Second, as it becomes okay not to do well in mathematics, it becomes okay not to do well in other subjects. As children develop beliefs about themselves and the world, if they feel like they are not good enough, that assumption will begin spilling into other subjects.

Math anxiety seeps into your child's entire education. Once there's a gap in your child's math understanding, math anxiety builds due to that gap. If your child doesn't understand the teacher during a math lesson, they just assume they're not good at the subject.

Our society tells them it's okay if they are just "not a numbers person." Once it's okay not to be good at one subject, it's easy to have excuses for being deficient in other subjects, too.

YOUR CHILD'S EXPERIENCE AFTER KINDERGARTEN

Because math concepts build on each other, if your child doesn't understand math during their first year of school, it will be challenging to build ideas on top of that. From the numbers, the curriculum leads students to addition and subtraction.

Maybe they didn't really understand counting, but now they're on to addition and subtraction, employing memorization as a technique to pass. Once they get to multiplication, the children who were great at memorization *look* like they're doing well.

However, if they did not understand the ideas, to the student it sounds like that third-year biochem class after missing the first two years that we described earlier. The course sounds like English, but the student does not understand what is being said, and they are looking for the facts that would be on the quiz to ensure they can move on.

How could this possibly be happening right under the parent's nose? Because children who rely on memorization don't appear to have any issues with math. If a parent asks their child what 4×5 is and the child quickly answers with 20, the parent can check that off their mental list. There doesn't appear to be a need to dig any deeper. If the grades do not warrant holding a child back, the school may not flag a problem, and our natural inclination as parents is not to create one.

However, as a parent, if you put up four groups of five objects and the child has to count all of them to know there are twenty items there, rather than recognizing that four groups of five equal twenty, *then* you see there is a problem. Another way a parent may notice something is off is if students are struggling with word problems. Many parents come to Elephant Learning knowing or feeling that their children are not understanding the basic math facts.

Take yourself back to that spot. You're that third-grade student who's not understanding anything going on. What are you going to do? You're going to use the strategy you know will get you to the next level, and everyone will help you with that strategy because everyone—parents, teachers, administrators—is motivated to help you pass your test.

What happens to students when they get to middle school, where individualized resources are scarcer and memorization maybe doesn't work as well as it did when students were learning multiplication tables?

When students get to middle school, they transition from understanding fundamental ideas to using those ideas to create conversations within algebra. Once a student gets into algebra, if they don't understand the concept, it's game over. Memorization as a strategy does not work with algebra because the student has to follow the conversation in order to apply the techniques.

WHAT HAPPENS AFTER HIGH SCHOOL?

The large majority of children whose families can't afford to send them to preschool end up creating a pipeline of kids who aren't well-versed in algebra. Algebra is the language the rest of mathematics takes place within. In fact, after completing algebra, students typically will go on to geometry. Geometry requires algebra as a basis, and it is an application of algebra to shapes. Trigonometry is a study of cyclical functions (algebra). Calculus builds new ideas on top of algebra, while statistics is also firmly based on algebra.

In 2019, National Academy of Early Childhood Programs (NAECP) statistics showed that 75 percent of high school students were not proficient in high school mathematics. Typically, this is due to a misconception originating in algebra or earlier. Even at the university level, we see students coming into office hours with misconceptions about algebra.

Sixty-nine percent of STEM majors switch to a major with less mathematics. We tell children they can grow up to be anything they want, but to be honest, it was over in kindergarten due to a small math gap that would have been easily surmountable. Anything we can do to change that could empower people at an unimaginable scale.

What do you need to know in order to just be good at, say, computer programming? It's algebra, logic, and problem solving.

When we say that 75 percent of students have a deficiency in high school, think about the impact it would have if you opened up that 75 percent of the population to the jobs they were promised, the jobs they want. What if we were able to graduate more engineers and scientists, and the businesspeople and politicians of tomorrow could understand what they were saying?

What sort of impact would that have on our society and this planet?

PARENTAL GUIDANCE REQUIRED

When parents are involved with their child's education, the outcomes for the student are *always* better. Fortunately, there are a lot of tools out there to help you take control of your child's mathematics education.

Elephant Learning accurately evaluates students to determine their initial level of understanding. We are able to quickly determine which subject matter a student understands and does not understand and then build from that point on.

The app provides educational games for the kids while also providing parents with reports and information on how the app is teaching the topics. There is parental advice on how to take learning outside of the system through games at the students' level. For example, students could play board games while learning to count. We break it down for you, telling you how to help your child along every step of the way and showing you how to easily identify your child's misunderstandings.

The most valuable tool is being able to identify what the misconceptions are, and this works for students within our system as well as students who are not in our system. If you work with your child and they are not getting it, allow them to answer a question and then ask them why they think the answer they chose is correct. When articulated, the parent is easily able to determine what the misunderstanding is and either offer a hint or distinguish the ideas so the student gets that aha moment and has their misunderstanding addressed directly.

For example, a good friend of mine used this technique with his daughter. He asked her a question about how much older one of her cousins was compared to another. After trying this technique, he found that she had conflated the ideas of "taller" and "older" and that, by distinguishing them, she was now able to understand the question he was asking.

Elephant Learning's core mission is to empower children with mathematics. When we empower parents with tools to help their children, we empower children. That is partly why I am writing this book. I want any parent to have the ability to do what we do on their own.

Chapter Four

How Math Determines Your Child's Overall Success

The latest research shows several things. The first important result shows that the earlier you invest in your child's education, the higher the return on investment. Research also shows time and again that early math skills heavily determine your child's overall success. So what more can parents do to help develop their child's early math skills?

From reading and writing to overall comprehension and problem solving, math plays a huge role in your child's future. They don't even need to go on to a remotely math-heavy career to experience benefits.

THE UNEXPECTED INDUSTRIES THAT REQUIRE MATH SKILLS

Studies have come out recently showing how critical math skills are to the job market. The US Chamber of Commerce Foundation found the following:

- The fastest-growing, best-paying jobs are in STEM-related fields.
- Careers in banking and finance, information technology, healthcare, and construction are all STEM-related jobs predicted to grow to over nine million by 2022.
- There's an increase in demand for STEM skills in non-STEM fields.

PayScale's latest College Salary Report is especially eye-opening. **STEM grads have a mid-career average salary of $103,408.** This is a big deal considering the cost of higher education and student loans.

As you see the digitization of different industries, you can watch as the skills needed for success in these industries evolve to include more math.

Take, for example, marketing. It's now a completely different field than it was ten or twenty years ago. With the rise of new ways to measure marketing performance, the field increasingly relies on analytics, machine learning, data, and reporting.

It is not just about the money. Think about the psychological impact of wanting to be a doctor or physicist and switching to English because of math. The choice comes from a place of disempowerment; therefore, the result can only be disempowering.

Chapter Five

How to Evaluate Your Child's Math Skills Based on Language

In order to increase your child's math skills, you have to identify the starting point of your child's comprehension. There are strategic ways to do this to ensure you're not confusing your child even more.

Rather than evaluate your child's math skills by giving them a problem like "What is 5 + 4?" or "What is 7 × 9?" where they could have memorized the answer, try to evaluate their skills based on language. If your student doesn't understand what is happening with the symbols beyond what is on a piece of paper, they're not going to be able to apply the math to solve real-life problems. Because success and ease with teaching come from comprehension rather than proficiency, this will also ensure that your student is able to understand what you are saying when you introduce new ideas.

Ask your child some word questions to determine if they understand what you mean when talking about math. Here are some examples of how you can test their understanding of math concepts.

COUNTING

To test counting, there are several things you can do. If you want to see if your student has comprehension, you may try a producing exercise with them.

Ask them to give you eight things. If the student is able to pass over eight items, stopping at eight without help, then they are proficient at counting. If they continue, it's okay; let them continue to see how high they can get, but in the process, you can then see if they are proficient at "How many?"

Many parents play with their children in this way

already. You hold out eight fingers and ask the child, "How many fingers am I holding up?" If they're able to count and get the answer, that's good. If they say something like, "Five and three," then you'll need to continue easing the problems until you can find which numbers they understand and which they do not.

If they're showing that they can produce, you may move ahead a step further. Ask them, "If you had one more, how many would you have?" Remember, it's okay if they count to get the answer or use their fingers; they will develop proficiency at mental math in time. Until then, let them use the strategies they understand to solve the problems. The slowness of the solution should enable them to come up with better strategies.

SUBTRACTION AND ADDITION

When it comes to subtraction and addition, you could ask a question like, "If I had eight peaches, and someone came and took four away, how many would I have left?" You can do this with or without fingers.

Because addition and subtraction are two sides of

the same idea, you can alternate between them, such as by asking, "If I had fourteen ships, and someone brought me two more, how many would I have?"

Remember, it's okay for your child to count to get the answer or to use their fingers. It means they understand how to solve the problem, and that demonstrates they understand the language. That's good.

If they are not proficient, that is okay, too. Go to easier subjects and find out what they understand and do not understand. Catching them up will be fast and easy if you know you're starting in the right place.

MULTIPLICATION

When it comes to evaluating multiplication and knowing whether your child understands the language and concept or has just memorized their multiplication tables, we use groupings by rows or collections.

You may arrange six rows of seven items and ask,

"How many are there?" If your child is counting to get the answer and they've already memorized their times tables, then this may be an indication of a problem. You may quickly be able to catch them up if you say something like "What is 6×7?" after they count to answer.

A lot of times, when students who have memorized their multiplication tables make this real-life connection of what multiplication actually means, they're able to get the concept. They glide through multiplication moving forward.

FRACTIONS, DECIMALS, AND PERCENTAGES

These three math topics are all representations of proportionality. Students who are having issues understanding the idea of proportionality will have issues with all three.

They are best evaluated visually. The easiest way to understand proportionality is through fractions, and the best way to do this in real life is through measurements—anything that requires a measuring cup, ruler, or measuring tape. By working with your student on a project like this, you will quickly see what they do and do not understand. When they do not understand, take note and back off; when they do understand, you may give them more challenging questions.

When it comes to fractions, it's important for children to understand that fractions, decimals, and percentages are all representations of the same idea: proportionality. It's just a quirk of human language that we've agreed upon these three different ways to represent the same idea. No one way is more correct than another. It's important to say this to a student. It is often overlooked in the classroom and ends up contributing to anxiety around these topics.

WHAT TO DO WHEN YOUR CHILD DOESN'T UNDERSTAND

What can you do if you're working with a student, but you're seeing that they don't understand something? It's pretty simple actually.

Let them answer the question incorrectly. Ask them why they think it is the correct answer. Based on their reasoning, you should be able to see exactly what the child does not understand. Then either give them a hint that allows them to have that aha moment or clarify the idea that is misunderstood.

This is the process we encourage our parents to use when working with their child in the Elephant Learning app. Every parent we have spoken to who has taken the advice above has been able to coach their child into understanding—including my friend who discovered his daughter thought "older" meant "taller," which was causing the issue!

It's very important not to get frustrated if your child does not understand a concept. This is where Elephant Learning really excels. We help you find your student's level without the exhaustion. The computer has infinite patience.

Please don't keep pounding away at the same concepts in the same manner if your child isn't getting it. It could be that they do not understand something more fundamental, and that is why they are not understanding the explanation or the hint. It is extremely common for the student to still be struggling with an idea that came before what they are working on now.

Sometimes it's best to take a break. Go online and see how other people are teaching these concepts; you can even look at how the Elephant Learning app is teaching a concept and replicate our methods.

BUILDING A SOLID MATH FOUNDATION

Evaluating your child's math skills is so much more than just giving them a math sheet filled with problems or looking at how well they're doing in class. It's all about ensuring they have a strong math foundation that holds up over time as they move into harder and harder concepts. Evaluating their comprehension based on the language surrounding math makes building this foundation that much easier so you can move into teaching math using a time-proven, three-step method.

Chapter Six

The Three-Step Method to Teaching Math

—

If you want to help your child learn new math concepts, then effective teaching and communication methods are your best tools. The first goal, though, is understanding your child's level. If you're not working with your student at their level, then it will be difficult for them to understand you, which will inhibit their learning.

The following three-step process to teach math effectively is what we use at Elephant Learning, and it is something you can use in your own work with your child. It includes defining an idea, determining if a child recognizes a definition, and then allowing the child to produce the idea in order to demonstrate their comprehension.

For example:

- You show your child three objects and say, "This is three." This is **defining**.
- You hold up three objects and ask your child, "How many do I have?" If they respond with "Three," you know they are **recognizing**.
- You ask your child to give you three objects, and they do so correctly. Now they are **producing** the idea in order to solve the problem. This is ultimately **comprehension**.

This last step is the proverbial check mark. If your child can do this, then they truly understand the math concept. This is not just true for mathematics—it is true for words in general. Think back to when you taught your child about colors. At first, you showed them red things, labeling them. Then you may have started asking "What color is this truck?" or "Could you give me the red truck?" Then, suddenly, your child was identifying red things out in the world. That is when we as parents know for sure that they understand the idea of red. The same is true of mathematics.

This three-step process can be used with all concepts that you would want to teach your child, including counting, subtraction, multiplication, and fractions.

1. DEFINE THE IDEA

Defining a math concept for your child is where the primary instruction is going to occur. You're going to show them an idea, have them exhibit the idea, and then label it with the name of the concept. Let's go back to a previous analogy we've made regarding how to teach a child their colors.

You can't just verbally tell a child what the color red is and expect them to understand it. What you do is you show a child red objects and then label them as red so they can then recognize the color at a later time. Similarly, you have to show a child a math concept and then label it, therefore defining the idea, before they can truly comprehend the concept. You can't describe to a child what addition is if you're not doing it.

Defining a math concept with a child isn't as complex as it sounds. Ask the child to give you five things and then four more things. Now how many do you have? Nine. That's addition. After the child counts to get the answer, let them know, "That's right! Four added to five is nine."

2. RECOGNIZE THE DEFINITION

Can your child accurately give you four building blocks, then five more building blocks, and then tell you how many they have total? If so, they recognize addition. At first, your child is likely going to count all nine building blocks in order to give you an answer, and that's okay.

But at some point, you want them to move to a more advanced counting strategy. They should be able to "count on," or start at five and then count on to nine, without having to count all the building blocks over again. The Elephant Learning app accomplishes this by hiding the first quantity of items from the app user so they're forced to start at their original number (five) and add on the four items they can actually see to get to the correct answer of nine.

3. DEMONSTRATE COMPREHENSION

Once your child is confidently and correctly using the concept to solve problems, they are demonstrating comprehension. For counting, this would be if they are able to produce, for example, seven objects and stop at seven. For addition and subtraction, that is identifying that addition or subtraction would answer a word problem or a real-life problem. For multiplication, that is using it as a tool to solve problems with grouping or arrays.

HOW TO DEAL WITH PITFALLS AND PROBLEMS

If you're working with your child and they're not demonstrating the ability to recognize the definition, it's just a matter of going back to the definition and explaining it again. This is where you don't want to get frustrated. You want to remain calm and patient, but don't repeat yourself too much. If you and your child are both frustrated because they're not getting the answers right, it will only lead to math anxiety and avoidance of math on their part down the road.

Maybe it's not a matter of your child not understanding, but rather you're not teaching the definition properly, at which point it might be time to consult Elephant Learning or another online resource to see how best to teach these definitions.

TEACHING MATH EFFECTIVELY

Teaching math effectively is so much more involved than asking them to memorize some multiplication tables. It requires passing on the experience of math concepts and ensuring a child truly comprehends them before going on to the next one. However, with a little bit of work and a lot of patience, parents can teach their children math in a way that sets them up for future success.

Chapter Seven

How to Gamify Your Math Lessons

———

Gamification is finding a way to make math fun for kids, making math play rather than work. The gamification of math lessons happens when you identify where math occurs in either actual games or in the world around you. It's fairly simple to do and can make all the difference when it comes to your child's relationship with math.

In this chapter, we'll look at what gamification is, why it matters, and how to put this concept to use in helping your child succeed in math.

GAMES WITH MATH VERSUS GAMIFICATION OF MATH

The concept of "gamification" can be difficult for some parents to grasp, often because of the influx of math apps and programs out there that are, in reality, just video games with math problems dropped in. This often isn't helpful to children struggling in math because the math part of the game is still a chore. It's something to just get through before you get back to the game. This is not gamifying math; they're just games with some math sprinkled in.

Making math fun isn't a bad thing, though. It just needs to happen correctly. When you "gamify" math, you're giving your child a fun math experience that keeps math as the focus. Elephant Learning's approach is to use our proven effective math curriculum as the foundation and build games out of the math rather than the other way around. You have to make the math itself enjoyable, rather than disguising math with fun from unrelated sources.

HOW TO GAMIFY MATH LESSONS IN REAL LIFE

Making math fun for your child within the confines

of your everyday world is easy. Let's say you're walking down the sidewalk with your child, and they say, "Oh, there's a train." That's an opportunity for you to ask how many train cars they can see. How many engines are on the train?

Even if it's just their toys sitting out on the floor, you could ask them, "Can you give me three toy dogs right now?" Then your child has to identify what's a dog, what's not a dog, and how many of them equal three.

Take whatever your child can identify and formulate a math lesson on their level.

HOW TO GAMIFY MATH LESSONS USING BOARD GAMES

Board games are an excellent way to make math fun for your child. There are lots of ways they can practice math skills during a family game night. For example, they have to roll the die, they have to identify the numbers on the die, then they have to produce that number of spaces on the board.

When using board games to gamify math lessons,

it's important that the game not be beyond your child's level of comprehension. If you're playing the game with them, it's important that you understand what's beyond their level and then do those parts of the game for them. You don't want to ask them to do anything beyond their level because that can cause frustration, and your child will no longer enjoy the experience (thus defeating the purpose of gamifying your math lesson).

Some board games that might be a good fit for your child's developing math skills could include Candyland or Chutes and Ladders. Both involve simple counting. If your child is beyond counting and moving on to other math skills, board games like Monopoly or The Game of Life could be more appropriate.

PITFALLS AND PROBLEMS

When gamifying a math lesson, remember that you're working with a human being. You wouldn't go to your job and start telling people they're blatantly wrong. Similarly, you can't do that with your toddler or preschooler. You can't say, "Oh, you're wrong. Why don't you get this?" Your child doesn't know *why* they don't get a concept. What are they going to say to you? Questions like these only lead to tears at math time—this isn't making the math experience fun at all!

Instead of telling your child they're wrong and asking why they don't understand, you want to ask them why they think their answer is right.

"Oh, you think five plus four is ten? Why do you think that?"

When your child tells you why, listen to the answer and do not try to correct them while they explain. You'll be able to realize exactly what the gap in understanding is. It is typically easy to either help them through defining or give them a hint that helps them figure out the correct answer. The hint is the preferred method because when a student gets an aha moment from solving a puzzle in real life or a game, they own the win and they build intuition.

BRINGING MATH TO LIFE FOR YOUR CHILD

Take math out of the classroom and bring it to life in a tangible, enjoyable way. Gamifying your math lessons, whether using an actual game or real-life scenarios, is a great method of making math fun for your child, not just work.

Chapter Eight

The Early Years

TEACHING YOUNG CHILDREN MATH CONCEPTS

———

Studies show that early math skills have far-reaching benefits beyond just school performance, so naturally you want to teach your child math concepts early to give them the best edge throughout their life and career.

But when do the "early years" begin? How early is too early? And for that matter, where do you start when the time comes? Here is everything you need to know about teaching early math, from understanding when a child is ready to learn to tips for teaching foundational math concepts.

IS MY CHILD READY TO LEARN MATH CONCEPTS?

Everything depends on the student. We've had some two-year-olds in the Elephant Learning system who have thrived, but we've also had two-year-olds in the system who just weren't ready yet.

You have to judge your child's readiness and honestly ask yourself if they are prepared for this step. These are crucial years for a child. In some cases, they're still learning to speak. Can they even say numbers? What's the point of asking them "How many?" in a math problem if they can't even articulate an answer?

Ask yourself:

- Can your child say numbers out loud?
- Can your child see numbers as numerals and then say them?
- Can your child begin counting?

If yes, great. Your child is likely ready to learn math concepts. But if not, no worries. There's honestly no hurry and here's why: it's very common for us to have students in the Elephant Learning system at ages four or five years who start at the very basic counting skills but then move on to multiplication and more difficult concepts in the span of three to six months.

THE FIRST STEP: IDENTIFYING WHAT YOUR CHILD ALREADY KNOWS

Regardless of your child's age, when you want to begin teaching them a math concept, you have to identify the starting point of their comprehension. This way, you can ensure you're not wasting your time or confusing your child.

Use language to find out what level your child is at. Which words do they understand and not understand?

If you're teaching your child during their early years, around the toddler age, it's likely that if they're familiar with *any* math concept, it's going to be counting (and if they're not, as stated above, that's fine too—there's more on how to introduce them to counting later in this chapter).

How can you test your child's counting abilities?

First, ask them how many items are in a group. Can they count them?

Then take the evaluation a step further and ask your child to produce, rather than just counting. Have them separate out a certain number of objects from a larger group of objects. This tests whether or not your child further comprehends the concept of counting and can stop counting once they reach the desired number.

Finally, you can then see if your child is able to "count on" by asking them to start counting beyond a number other than zero. For example, you could ask, "How many would I have if I had two more than the eight you just gave me?" If they can "count on," then they'll start counting at nine. If not, they'll begin counting the total number of objects all over again from one.

Does your child know how to count? If so, then you can move on to other concepts. If not, here's how to get them started.

TEACHING YOUR CHILD TO COUNT

When teaching children to count, it's all about **definition**, **recognition**, and **production**.

The first step is being able to define a number. Can your child recognize a numeral? Can they say the word?

Show your child a singular object, call it "one," and show them the numeral "1." Use the language in context, and help them learn to define numbers. If they've mastered this first definition phase, they'll be able to say the word "one," recognize the symbol "1," say "one" when they see the symbol, and recognize one object.

Once you're confident your child understands the definition of "one," then you can test their recognition. This test is as simple as holding up fingers and asking, "How many?" You can ask the same question throughout your daily life. If you see them counting to get the answer, they're showing recognition.

After your child can recognize numbers and counting, move on to production. Ask them to give you a certain number of objects. If they're able to hand you that number of objects and stop there, then they're able to produce, and they've mastered counting.

In the Elephant Learning system, we start by teaching children to count from one to five. After they get to a certain level of recognition, we start teaching them to identify five through ten while they're learning to produce in quantities of one to five. After they can recognize those numbers, we move on to double digits: eleven, twelve, and thirteen. Then we can start to establish the rest of the teens and up to twenty.

STAY CALM THROUGH THE LEARNING PROCESS

The one thing you want to make sure you do during this process is stay calm. Keep in mind children at this age are prone to forget math concepts from day to day. If, for example, your child learns how to "count on" on Tuesday but then can't recall how to do so on Wednesday, it's no big deal.

Additionally, children of this age often have issues

with their attention spans. Just be aware of your child's attention span, and don't try to push them beyond it. Hold their attention as long as possible, but don't force anything—that can just make them tired and cranky.

Remember, this is hard work for children. Counting on your fingers seems easy to me and you, but for a child, it's a challenge.

The key point is to understand where your child is and work with them where they are. If you start to challenge them and they don't respond positively, just take things back a step and keep practicing. The more you practice a math skill, the faster they'll learn it.

THE RESULT?

If you can do all of the above and get your child counting to twenty, then they're entering kindergarten ahead of their peers, and statistically speaking, they have a good chance at going to college and being prepared for STEM fields.

Chapter Nine

Elementary Mathematics

Early elementary mathematics spans the ages of six to eight years old—roughly kindergarten through second grade. Though mathematics curriculum varies from state to state and school to school, kindergarten through second grade is where children learn the fundamentals of math: counting, comparisons, addition, and subtraction. Children are also introduced to skip counting and the number line, two strategies that set the foundation for later elementary math.

As a parent, you already know how important it is for your child to grasp these early concepts, and you may be looking for a math app to help them excel. Let's take a look at how the Elephant Learning app teaches each of these early elementary math concepts.

COUNTING AND COMPARISONS

In early elementary education, the first concepts that we work with are counting and comparisons—that is, quantity comparisons versus what's bigger and smaller. We might show a child an image of four objects and an image with twelve objects and ask them to identify which has more or fewer. It's important for children to know the difference because it sets the stage for addition and subtraction.

ADDITION AND SUBTRACTION

In the Elephant Learning app, there's a seamless transition from counting and comparison to addition and subtraction. This is actually why many of our young students are doing so well. We're simply walking them logically through what you would want to teach a kid to get to the very next baby step.

The question, "Can you give me five items?" incrementally morphs into "Give me five items. Give me one item. How many do I have now?" or "If I have five things and someone takes one away, how many do I have now?" The child learns the order of the numbers, which becomes addition and subtraction. This helps establish the order of the numbers in their mind, which helps them to develop numeracy—the ability to understand and work with numbers.

The Elephant Learning app addresses these concepts from numerous angles. One question might ask, "A farmer had fifteen carrots and gave three to his horse. How many does he have left?" We then approach the problem from a different angle and ask, "A farmer had fifteen carrots and gave some to his horse. Now he has twelve carrots. How many carrots did he give to his horse?"

By approaching the same idea from multiple angles, we help the student understand all of the language that may be used, as well as having them solve the same problem from a different angle. When they do, they are not only showing they are proficient, but they are also understanding the idea on a more intuitive level.

SKIP COUNTING

The other math skill that children work on during the early elementary years is skip counting—two, four, six, eight, and so on. The idea is for the child to start to see the grouping. Skip counting really is the precursor to multiplication, and the more advanced skip counting questions at school and in the Elephant Learning app look a lot like multiplication questions.

THE NUMBER LINE

Building numeracy requires students to have an understanding of all representations of numbers. We work on numeracy using objects, though at some point it is good to abstract to a number line. This helps students see the numbers placed out sequentially in order on a horizontal line. It allows them to approach addition and subtraction from a different angle and allows us to determine their proficiency with numeracy.

For instance, our app might show a child a number line and ask, "Where's 17 and where's 71 on the line?" We ask this question on a number line going from 0 to 100. If the student places 17 near 71, then we know they are having an issue understanding two-digit numbers. However, if they are answering correctly, we know they have mastery of these ideas.

ELEPHANT LEARNING ACCELERATES EARLY ELEMENTARY MATH

Early elementary mathematics focuses on the fundamentals: counting and comparisons, addition and subtraction, skip counting, and the number line. Using the Elephant Learning app, children can learn these early elementary mathematical concepts in a matter of two to three weeks, as compared to two years of the standard school curriculum.

The best part is that once a child has the understanding, a teacher can't take it away from them. Even if there's a difference between the way that the school teaches a concept and the way the child learns with Elephant Learning, parents can reconcile the information because the concepts are solid. Mastering these skills sets the foundation for the years ahead when children will tackle multiplication, division, fractions, decimals, percentages, and more.

Chapter Ten

Later Elementary Math Concepts and Strategies

When a child enters the later elementary stage of their education (for our purposes, consider it third through sixth grade), they'll likely be learning multiplication, division, fractions, percentages, and decimals. Even though children have moved on from the early elementary topics of addition and subtraction (which they'll need to master if they want to succeed at later elementary topics—things get complex quickly), parents will find that moving from addition and subtraction to multiplication and division is a natural step. After all, multiplication and division are essentially repetitive addition and solve problems that use grouping or splitting. A multiplication problem might represent four groups of six items, and we know that four groups of six items sums up to twenty-four total items.

The Elephant Learning app teaches these types of concepts through a combination of hiding items (so your child can't use counting to solve a multiplication problem), timing, and other strategies to develop children's math skills.

How can you introduce and reinforce these concepts at home and throughout everyday life? What are some of the best ways to make these concepts "click" for your child? Here are a few strategies to set your child up for success in these areas.

MULTIPLICATION BEYOND MEMORIZATION

Most students learn to multiply in school by memorizing their multiplication tables. There's nothing wrong with memorizing multiplication tables, but a child must know what the multiplication tables *mean*. If they're multiplying seven by six, they need to have that picture in the back of their head of six groups of seven or seven groups of six. If not, they don't have a true understanding of what multiplication actually is, and it won't serve them later in life.

Take, for example, a child who knows that five times four is twenty. She can solve the multiplication problem with ease. But then she looks at a real-life problem she could solve with multiplication. Maybe she's looking at a group of objects separated into four groups of five. If she starts counting the objects one by one, this tells you she doesn't understand what multiplication is and how to use it in everyday life. She just memorized the multiplication table, which is absolutely useless to her in the real world.

When working with your child on multiplication, make sure they understand the *meaning* of multiplication. Go beyond mere memorization.

DIVISION: NOT SO DIFFERENT

Just like addition and subtraction are the same topics essentially—just the inverse functions of each other—so are multiplication and division. Traditionally, children are taught multiplication first, then division, but it doesn't have to be this way. You could teach division first, then multiplication, and have the same success. Here's why.

Let's say A divided by B equals C. That means that A equals B times C. Division and multiplication are parts of the same equation. Which operation you use to solve a problem depends on what you're looking for.

If your child is successful in multiplication, division should be a natural step forward if you introduce it to them as the other side of multiplication. It's nothing new, foreign, or scary. It's something they've already encountered and don't need to fear.

THE SECRET ABOUT FRACTIONS, DECIMALS, AND PERCENTAGES

Division, fractions, decimals, and percentages are all part of the same concept—proportions—so why are they taught separately? The Elephant Learning app introduces fractions at the same time a child is learning multiplication and division. After all, a fraction is division; one over four literally means one piece of a whole divided by four. Traditional instruction introduces division, fractions, decimals, and percentages as separate concepts.

No method of depicting a proportion is more correct than another, though depending on the context, people typically use one over the other. For example, money makes the most sense with decimals, financials with percentages, and projects that require measurements with fractions.

When parents themselves realize that fractions, decimals, and percentages are all different ways of saying the same thing, it's like a light bulb goes off. Many people go their entire lives not realizing it! But once they do, it seems so obvious.

Parents can help their child come to understand this by not only using the language surrounding these concepts in everyday life but by showing them the concepts in everyday life, too. Maybe you bake a cake using measuring cups, or you work on a DIY project that requires measuring tape. Suddenly, it becomes very real to your child what a quarter of an inch is.

Fractions are often easiest to start with, as out of all

these concepts they look and feel most like division. Then you can step your child up to decimals. They may look intimidating, but all you're really doing is creating a fraction where the denominator is 100 (or one plus the same number of zeroes for however many decimal places you might have). So 0.22 is simply 22 over 100, or 0.122 is simply 122 over 1,000. Once they grasp this concept, you can move on to percentages. The percentage is just like a fraction, but the denominator will always be 100; 50 percent is 0.50 is 50/100.

All three represent exactly the same idea, just in different languages.

NORMALIZING MATH LANGUAGE

Why is there such a stigma around math terminology? Research shows that children as young as four years old exhibit the concept of division all the time. Think about how they divide up their toys for a tea party or how they divide up a snack. They usually have an idea of what a half or a fourth means at that age. They're using division, but we don't label it as such in normal conversation.

We need to start integrating "formal" math language into everyday talk. Using math words and terms around your child introduces them to the concepts as soon as they start learning language. It makes them more comfortable and confident with math as they get older. Do math out loud in front of them. Count on your fingers. Talk through an everyday math problem with them. Don't be afraid of looking stupid just because society says that counting on your fingers makes you look dumb.

NEXT STEPS: WHAT HAPPENS AFTER LATE-ELEMENTARY EDUCATION?

After late-elementary education, your child will get into algebra. They'll need to understand all the previously discussed concepts in order to succeed. While it is possible to teach and reinforce these concepts to your child at home, it can be a lot of hard work and very time consuming. Elephant Learning can help.

Chapter Eleven

What Parents Need to Know about Math Curriculum in Algebra and Beyond

---

When students move into algebra, they begin using mathematics to have conversations.

Prior to algebra, everything we do is really definition. It's teaching children what numbers, addition, and subtraction are and how to think about them from different perspectives via the number line, groupings, quantities, fractions, and decimals. Then they transition into algebra and start using formal mathematical language. They begin writing in the language of math, the language they've been learning this entire time.

Sadly, this entire process often happens without children even knowing it. The algebra teacher doesn't explain that everything they need to understand has been a prerequisite to understanding the conversations they will have. This can cause a lot of stress for your child both at school and at home.

Luckily, you can work with your child to help them develop their language, just like we did for early elementary and late-elementary concepts.

Here are two key places students get confused when faced with algebra for the first time and how you can help them overcome these issues. This is by no means an exhaustive list, but it's a great starting point for ensuring your student is ready for algebra and beyond.

STUMBLING BLOCK #1: THE EQUAL SIGN

One place where children have a lot of issues with algebra is the equal sign. The equal sign basically means that the quantities on both sides of the equation are the same. We notice this when students do not understand, for example, "Why are we subtracting five from both sides?"

A gamification of the idea of the equal sign typically uses a balance and changing quantities on either side to show the relationship between more, equal, and less. However, at this age, you can just work with students on the definition so they can understand it. Playing with ideas involves testing them to see if they can identify when the symbols $>$, $=$, and $<$ are used in statements that are true or false. For example: $5 = 5$ (true), $4 = 5$ (false), or $4 < 5$ (true).

Once the student is able to communicate the above ideas, a parent or teacher can begin to teach more complicated language. For example, an equation is two expressions that are related by an equal sign. An expression is any statement that you can make in mathematics such as $5x + 5$, $5 + 4$, and so forth.

Because math jargon quickly builds upon itself, it is very important to have understanding every step of the way because otherwise it is very easy to lose students when statements are made about more complicated objects.

STUMBLING BLOCK #2: MEMORIZATION

Algebra, as a practice and division of mathematics, deals with abstraction. The conversations and ideas are more conceptual (but no less precise) than earlier mathematics. Rules of thumb that previously could help students achieve success no longer apply.

Because of this, another stumbling block is that many children at this stage are still relying on memorization skills that they may have picked up when learning their multiplication tables. They just want to memorize the steps to solve an equation so they can pass a test.

Unfortunately, if you're showing a child how to solve a problem and they attempt to memorize the steps instead of understanding why the strategies work, it will be difficult for them to achieve the goal of passing the test. Algebra is an exercise in problem solving, and it uses all of the language that came prior in elementary school. An equation such as $5x + 4 = 9$ requires both multiplication and addition and needs the student to understand the quantity on the left of the equal sign is the same as the quantity on the right in order for the student to be able to even begin to approach solving the problem.

HELPING YOUR CHILD OVERCOME STUMBLING BLOCKS TO SUCCEED IN ALGEBRA

Helping your child succeed in algebra starts with helping them understand some definitions. For example, look at all the basic symbols you'll be using, such as exponents, the equal sign, or the greater-than sign. The Elephant Learning app does a very thorough and rigorous job of defining these concepts to eliminate any of your child's common misconceptions.

After showing a child the definitions, we then test their knowledge of the definitions by presenting them with true or false statements. As they begin to develop an idea of what true and false means in terms of algebra, they begin to build their logic skills. The more students develop their logic, the more intuition they'll have when it comes to problem-solving skills, taking their math ability to an entirely new level.

THREE STEPS TO SUCCESS

At the end of the day, algebra comes down to these three steps: define, recognize, and produce. No matter if your child is in middle school or a doctorate math program, it's all about defining (can you understand it?), recognizing (can you identify it?), and producing (can you use it to produce results or new research?). If you can help your child with these three aspects of algebra at home, they'll be better set up for success in the classroom and the future.

Chapter Twelve

It's about More than Just Math: Fear, Growth, and Adaptation

Math anxiety—a fear of getting math concepts and problems wrong, and the resulting avoidance of math because of that—is something I've seen many times over my life, and not just in children. It's just as prevalent in adults, and despite my doctorate degree in math, I experienced math anxiety as a child, too.

While some children allowed their math anxiety to grow into a lifelong avoidance of math, mine fueled my competitive spirit and led me to push ahead of my peers, learning advanced math concepts even when I wasn't able to get into the advanced math classes my middle school offered.

That's a big question mark in the math anxiety experience and one that can greatly impact your child's future. Will they choose to avoid math for life? Or will they use math to their advantage in not just their elementary education but also their higher education and subsequent career? Will they have a growth mindset or a fear mindset? Will they avoid the concepts they fear or use their fear of math to get better at it? In many cases, these are the big questions to ask, not simply whether or not your child has math anxiety.

See, most of us have some sort of anxiety around math or another subject. The anxiety might not even be about the math *per se*. Instead, it's the anxiety around being perceived as a bad student or as "stupid." It just so happens that many people don't learn math easily via the curriculum used in most schools, and our society in general tells us it's okay to not be a "numbers person"—so math anxiety continues.

But if your child latches on to that growth mindset

and overcomes the fear of math, the opportunities are endless.

THE POWER OF A GROWTH MINDSET

When we encourage our children to pursue math and overcome their anxiety—instead of telling them being "not a numbers person" is perfectly fine—everyone benefits. It's not just about your child's elementary school grades. Beyond that, your child and their peers could be the catalysts for a better future for the world.

Math skills are analytic and reasoning skills. Students who do well in math usually do well in everything else. Studies have proven time and time again that children who do well in math early on do better in all their subjects later. A math-literate society is a more successful one.

A math-literate society can produce more scientists, technologists, engineers, and others who are equipped to solve the world's problems. Math-literate entrepreneurs, politicians, and creatives add their own value when they're able to discuss the world's issues with math-focused professionals.

BUT WHAT IF MY CHILD SIMPLY CAN'T DO THAT?

Some parents worry that their children are simply incapable of learning advanced math concepts, or even basic math concepts, due to a learning disability. But I feel that nearly every child can learn math regardless, and here's why.

Every student seems to have the capacity for learning language. At Elephant Learning, we work with math as a language, and if your student, regardless of learning impairment, is able to speak and understand language, then our system should be able to work for them (as it's language based).

BEYOND MATH

Similarly, just as we use language-learning methods to teach math within the Elephant Learning app, the same methods we use to teach math (and the same teaching methods discussed throughout the Elephant Learning blog) are applicable to any subject.

For example, one of the key ways we tell parents to help their child overcome math misunderstandings is, when a child gets a math problem wrong, instead of telling them the answer is wrong, ask them *why* they think that's the right answer. When a child explains, the parent can generally pinpoint why they're getting the concept wrong and remedy the situation. This same practice can be used when helping a child learn anything.

AWARENESS AND ADAPTATION

Through methods like this, learned through math, children can then learn to be aware of their obstacles and adapt to overcome them. But first, you as their parents have to be aware of math anxiety. Once they're empowered and go on to become aware of obstacles on their own, the sky's the limit. They can encounter a problem, and rather than letting their anxiety tell them to head in the other direction, they can devise a solution.

The empowerment children need in order to do so is possible through the Elephant Learning app and through working with your child hands-on on a regular basis and getting involved directly with their education.

With awareness and adaptation, your child can

accomplish anything—from overcoming their math anxiety to changing the world.

Chapter Thirteen

Children Are Empowered through Understanding

———

Many of us have been told we're "just not a numbers person." Half of all Americans report math anxiety. There's no such thing as being "not a numbers person," though, and it's never too late to learn math. Mathematics is really about learning a jargon, a vocabulary fundamental to humanity, and anyone can do it.

Unfortunately for a lot of people (the half of all Americans reporting math anxiety), the confidence they have when approaching math is the same confidence they have when approaching other areas, especially academics. Research has shown that preschool math scores predict fifth grade overall scores. Children who do more mathematics at a young age are better readers, writers, speakers, and problem solvers.

Mathematics is special in that it teaches more than just how to solve an equation. It exercises the mind so that when you're successful in math, you can be successful in other subjects.

How does success in math impact your overall success in life?

THE TWO REASONS SUCCESS IN MATH EMPOWERS CHILDREN FOR LIFE

There are two ways success in math impacts overall success in life.

1. WHEN YOU'RE PERFORMING MATHEMATICS, YOU'RE DEVELOPING MENTAL TOOLS

Think about it this way: when you start using a physical tool, like a screwdriver or a hammer, you're not automatically the most proficient carpenter there is, regardless of whether or not you have carpenter tools at your disposal.

But the more you use those physical tools—or, in math's case, mental tools—the better you get at using them. When you're performing mathematics, you're practicing using these mental tools that you can then use in other areas or situations.

For example, take one of the more basic tools that children learn when developing math skills (and a tool that some adults still cannot master!): the ability to solve a math problem in your head, to work with numbers without seeing them written in front of you. When you can accurately solve that difficult math problem in your head, no paper or pen required, think of how confident you'll feel. Which leads us to the second reason success in math empowers children for life.

2. THE MORE YOU DEVELOP THESE MENTAL TOOLS, THE MORE CONFIDENT YOU BECOME

The more you develop the mental tools, the more comfortable and confident you'll feel using them.

Children are more geared toward learning language than math, but if you teach them about the language *of* math (as the Elephant Learning app does), then they'll take that language and begin using the associated mental tools in everyday life. We constantly hear from parents that after their children have used the Elephant Learning app for a while, they become so comfortable and confident using these tools that they start using math to solve problems in their everyday lives. It's no longer just an academic chore—it's a real-life problem-solving tool that they feel empowered to use because the Elephant Learning app helped them understand it.

SUCCESS IN MATH CAN BE HINDERED EARLY

Unfortunately, getting kids to develop these mental tools and reap the confidence that follows is easier said than done.

It all comes down to much of society's attitude regarding math. We mentioned above how many people think they're "not a numbers person," but where did they develop that attitude? Does it actually reflect their ability to do math? Of course not.

That attitude develops in the classroom. As soon as a child struggles with math, they're given the excuse (from teachers, parents, or their peers) that it's okay because they're "not a numbers person." Once they hear that, it's an excuse not to try to get any better at math; it's an excuse not to practice using those mental tools. And if they never start using those mental tools, the confidence never develops.

Here's the truth: there's no such thing as being "not a numbers person." In reality, anyone can be a numbers person if they're willing to practice using the mental tools math requires.

THE RAMIFICATIONS OF NOT ACHIEVING EARLY SUCCESS IN MATH

Unfortunately, if a child is passed through the system like this and they never develop the mental tools that would make them confident, they may firmly believe that they're "not a numbers person." This may also lead to thinking it's okay to be "not a history person" or "not a literature person." Suddenly, that excuse has the potential to bleed into every other subject.

WHAT'S POSSIBLE IF WE HAVE A MORE MATH-LITERATE SOCIETY

But what if we had a more math-literate society, one filled with students who have developed those mental tools and confidence?

The United States actually had a heavily math-literate society not too long ago. It helped the Allies win World War II, took us to the moon, and led to the advent of the internet, just to name a few accomplishments. But we're not creating environments for those types of math-literate people to thrive anymore. Instead, we're 69 to 75 percent math illiterate at the high school level. Somewhere, there's a disconnect.

However, more and more, math is absolutely required for success in a growing number of fields. Beyond STEM fields, look at marketing. Once upon a time, marketing was an entirely creative field, but now it's completely data driven. Now if you want to go into a non-math-related field, you have to choose a humanities major, and statistically, those majors generally lead to lower-paying jobs. Unfortunately, because so many people are math illiterate, more and more people are entering the job market in lower-paying jobs that lead to more student debt and a lower earning cap overall.

Apart from the individual repercussions of math illiteracy, a math-literate society as a whole could offer worldwide benefits. If we produce more math-literate scientists, technologists, engineers, and mathematicians, they'd be solving the world's problems. At the same time, if we could create entrepreneurs and politicians who could also understand the math and what these math-focused professionals were saying, imagine what would be possible.

IT STARTS NOW

To achieve this kind of math-literate society, though, we have to start now—at your kitchen table, with your child. It requires tossing out the idea of being "not a numbers person." It means giving children the mental tools and confidence they need to suc-ceed, whether that success comes from you working one-on-one with them on a regular basis or using the Elephant Learning app.

Conclusion

———

Regardless of the experience you personally had with math in school, and even your relationship with math now, you can provide your child with a new approach to math—a new way to view math and the problems it presents, and a new way to tackle challenges, grow, and adapt.

This book was created to help parents help their children. It was created for the parents of perfectly average students who want to get ahead, as well as the parents with children struggling to stay afloat in math and, potentially, in other subjects. It was created for the parents who recognize the traditional way of learning math in the classroom isn't helping their child reach their full potential.

In order to help those parents and their children, I've laid out—as you've already seen over the course of this book—how to instruct children in math concepts from the nursery up. I broke down what's wrong with mere memorization and how kids can take a step back and view math as a language. I talked about the amazing potential for a math-literate society.

While the Elephant Learning app can certainly assist you in exploring these concepts further, hopefully this book alone empowered you as a parent, guardian, or educator to better instruct the children in your life. Maybe it even empowered you to look at math differently yourself. Maybe, for example, you will no longer approach balancing the checkbook with trepidation because of something you read in these pages.

But beyond laying out all these concepts, breaking down the facts, and empowering parents and children alike, as I put together this book, I realized Elephant Learning's approach does something more. We not only give parents and children the tools to succeed in math; we also give them tools to succeed in life.

See, math is just the beginning. When a child conquers learning anxiety and the fear of getting the wrong answer to math problems at homework time, that's the first step to conquering anxiety around any subject. It teaches them that they can overcome any challenge they encounter, even where there's a risk they might be perceived as "stupid" or "dumb." After all, isn't that what math anxiety is all about—the fear of being perceived as a bad student?

When a child can overcome this anxiety in both education and life, a world of opportunities opens up to them. They start to develop a growth mindset rather than a mindset of fear or avoidance. They not only approach their education differently; they also approach life differently. Armed with that growth mindset, they can do anything. They can change the world through their careers, their activism, their passions. And it all goes back to the child learning to approach math and their fear of math from a new perspective.

Is this a lot to ask of one book on math education? Is it too lofty to think that in these pages, parents might find the tools they need to empower their child for a lifetime of achievement?

If we said yes, well then, we'd likely be giving in to that anxious, avoidance-prone mindset. But saying no—that it's not too much to ask and it's not too lofty a goal—that's a growth mindset.

So if you've finished this book and feel as if you've been empowered to help your child see math—and the world!—differently, I've done my job. Now pass it on. Take the next steps. Whether it's subscribing to the Elephant Learning app to push your child even further, recommending this book to other parents or educators, or just making more time to explore education with your child, don't let the efforts stop here.

There's a big world of possibilities out there just waiting for our children. With the right tools and the right mindset—and a little math—they can conquer it.

About the Author

___

DR. ADITYA NAGRATH is on a mission to change the way the world teaches mathematics. He is the co-founder and chancellor of Elephant Learning Math Academy in Denver, Colorado, which uses proven techniques to transform children's mathematical learning. On average, students at the academy learn one and a half years' worth of math in three months just by using the system thirty minutes per week.

Dr. Nagrath holds a PhD in mathematics and computer science from the University of Denver. He also founded Elephant Head Software, where he led a team of engineers who brought more than thirty-five product lines to market between 2009 and 2016.

Made in United States
North Haven, CT
31 May 2024

53166860R00074